Tea Packaging Innovations

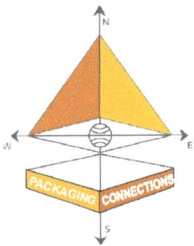

Sanex Packaging Connections Pvt. Ltd.
www.packagingconnections.com

Copyright

Published by :

Sanex Packaging Connections Pvt. Ltd.

An ISO 9001 : 2008 Certified Organisation

117, Suncity Tower, Sector-54

Golf Course Road, Gurgoan-122 002.

Tel : +91 124 4965770

Fax : + 91 124 41433951

e-mail : info@packagingconnections.com

Like us on Facebook : www.facebook.com/pconnection

TEA PACKAGING INNOVATIONS

List of Contributors

Team www.PackagingConnections.com by Sanex Packaging Connections Pvt Ltd

Sandeep Kumar Goyal, Founder & CEO
Amita Venkatesh Valleesha, Associate: Scientific Affairs & Consultancy
Chhavi Goel, Associate: Research & Business Consulting
Bhaskar Ch, Technology Advisor e-business
Sonu Sheoran, Associate Research & Technology

Table of Contents

Introduction

In our endeavour to get innovations which will take packaging to a different level, we have launched our first book on Product Packaging Innovations. These will be a series of books which will cover innovations in packaging for one product and each time a different product. We have started with Packaging Innovations in Tea and this covers different forms of tea as well as presentation of tea, with packaging designs, concepts, different packaging materials, patterns and many more.

Welcome to the world of packaging innovations! Find your innovation moment with our books.

Sandeep Kumar Goyal
Founder & CEO ,
www.PackagingConnections.com

TEA BAGS INNOVATION

TEA PACKAGING INNOVATIONS

Manufacturer/Designer

Sophie Pépin
Teapee, amerindian herbal teas

- This tea pack represents American nomadic lifestyle.

- Once opened these beautifully designed teepee packaging reveal tea bags which are formed in a similar teepee style shape."

- Each flavor has its own native American pattern which travels all the way around the packaging, serving as an opening mechanism.

- This tea pack is not only directly inspired by Teepees but also celebrates the skills of their legendary medical knowledge.

Manufacturer/Designer

Maja Matas, Kresimir Miloloza, and Jozo Matas

- The Christmas trees tea bags are complete one bag, when they are first pulled out of the box.

- Each tree contains two bags.

- They are then separated through perforation provided between tree and each half has a tea bag, creating the perfect tea-for-two.

- The Christmas tree halves sit gently on the side of the mug while the tea steeps.

- It encourages the sharing of a hot beverage with a dear person in the Christmas spirit.

Foto: Stefan Höderath

Foto: Stefan Höderath

Manufacturer/Designer

Janja Maidl

- The tea bag is put with the label on the cup edge and presents a strip that indicates the desired intensity of the tea.

- Each tea variety has a special paper strip which the tea climbs up as it steeps to show when it is ready.

- The principle is most beneficial with culinary and medical teas.

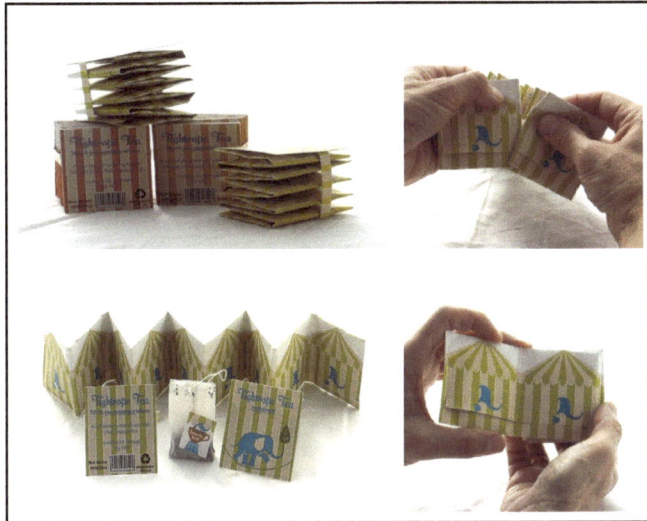

Manufacturer/Designer

Jondelle Watkins
Tightrope Tea

* Tightrope Tea is the result of a personal brief developed in response to excessive packaging of bag tea.

* The result of this is a compact, brightly colored tea packaging solution which reduces the amount of materials used during manufacturing.

* Tightrope Tea is a playful take on conventional teabag packaging, bringing the Colour, life and energy of the circus to the everyday cup of herbal tea.

TEA PACKAGING INNOVATIONS

Manufacturer/Designer

Yena Lee

- This is cleverly packaged as the butterfly is attached with thread of tea bag.

- Butterfly give creative edge to tea bag so as to be bought by the curious minds and impulse shoppers.

- In Packaging world quality products do sell well, but to give them that creative edge, they need to be packaged well. And this butter fly is bag is one of best examples.

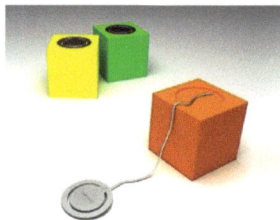

1. Lift up the clip.
2. Put the teabag in the clip.
3. Pull the clip to set the timer.
4. Get the teabag in a cup.
5. Teacube draws the teabag in 3 minutes.
6. Tea leaves are not brewed anymore.

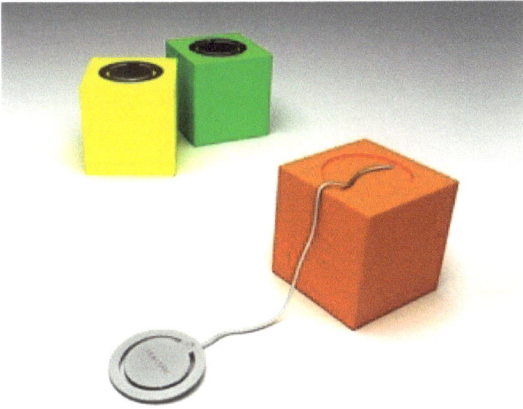

Manufacturer/Designer

Jieun Yang & Hanah Suh

• Tea Cube Makes Steeping Easy.

• Unlike instant coffee where we can just add and stir, making a good cup of tea is all about timing. If we leave the tea bag too long, it end up with a bitter, too-strong taste. Conversely, if we do not give it enough time to steep, the flavor won't be there.

• Solve this problem with the TeaCube.

• Simply attach the teabag to the clip, pull it to set the timer, and put the teabag in the cup. The TeaCube will automatically draw the teabag from the water after three minutes.

TEA PACKAGING INNOVATIONS

Manufacturer/Designer

Peter Hewitt

- This is a Tall polyhedral infuser tea packs, each with a disarmingly natural-looking leaf/sprout tag.

- A pleasing combination of geometric and organic: tall polyhedral infuser packs.

- Although the tall pyramidal shape is not a "close packing" polyhedron, the individual tea-bag/infusers do pack together in a variety of sculpturally intriguing way.

Information	**Tea bag package / 75 x 75 x 70**
Material	**Linen paper**
Feature	**Hexagon structure for flower tea, Octagon structure for leaf tea**
	People can save the environment with CUPTEA by reducing disposable tea package

Manufacturer/Designer

Designer Lee Seo-Jin
Origami Teacups

• The Cuptea brings the art of papercraft to beverage packaging.

• Two pleated Cuptea teacups have been conceived with the hexagonal one for flower tea And the chamfered square-based cup for loose leaf.

• Both are distributed flat and folded with herbs stored in the bottom, able to open and bloom into a small recyclable receptacle which simply requires hot water.

• The Cuptea concept allows the consumer to tote their own blend and collapsible mug in their purse or pocket.

• Cuptea concept makes the entire activity self-contained. Designer Lee Seo-Jin suggests amalgamating the infuser and the vessel into one.

TEA PACKAGING INNOVATIONS

Manufacturer/Designer

Developed by Lipton Tea

- Lipton Premium Pyramid Tea Bags.

- Lipton's new premium long-leaf tea products are being packaged in unique pyramid-shaped tea bags.

- This illustration is one of many that illustrate the tea blend, within the innovative tea bag, for the entire product line.

TEA PACKAGING INNOVATIONS

TEA PACKAGING INNOVATIONS

Manufacturer/Designer

**House Café in Istanbul,
Turkey**

- Creative cut-out images of people doing outdoor sports were attached to the strings of herbal tea bags at the House Café in Istanbul, Turkey.

- Cool tea packaging designed to associate herbal tea with a healthy lifestyle.

Manufacturer/Designer

**St. James's Teas (UK) ,
Teaosophy (U.S.) and
Adagio's (U.S.) pack**

- These innovative and high quality products have made tea packers think again about how they present the new-style tea bags to the retail and catering market.

- For use in the catering industry, St. James's Teas (UK) and Teaosophy (U.S.) pack each of their pyramids into an individual pyramid carton for a neat, stylish, cutting edge presentation.

- Adagio's (U.S.) 5 gram bag comes inside a flat, printed foil; té teas pack 15 bags to one flat-topped pyramid and a single bag in a mini version (for food service).

TEA PACKAGING INNOVATIONS

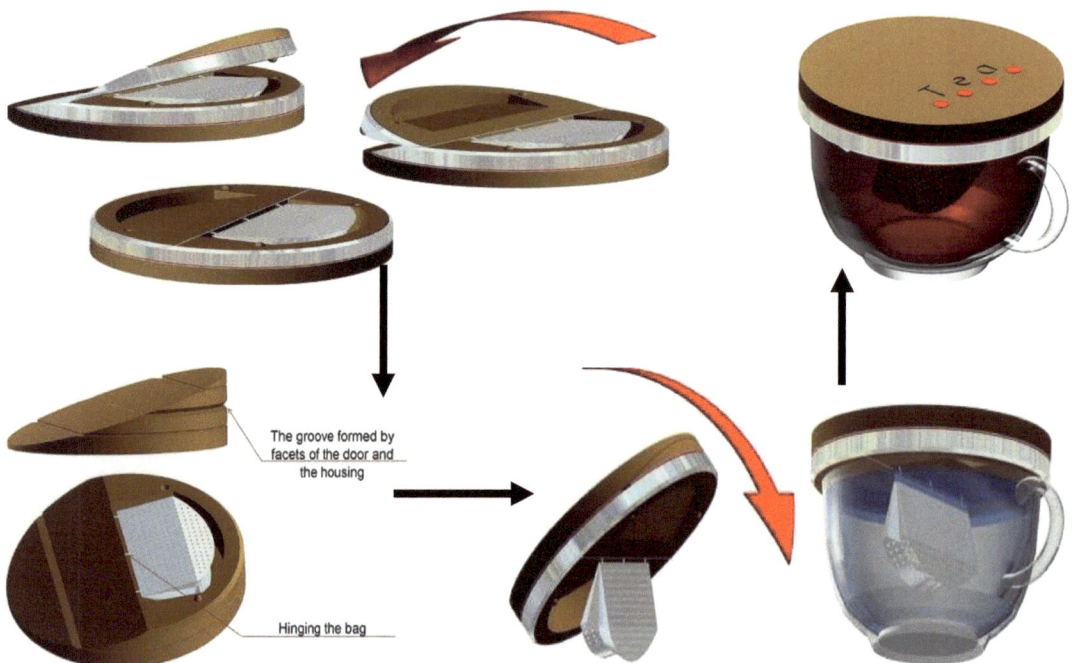

The groove formed by facets of the door and the housing

Hinging the bag

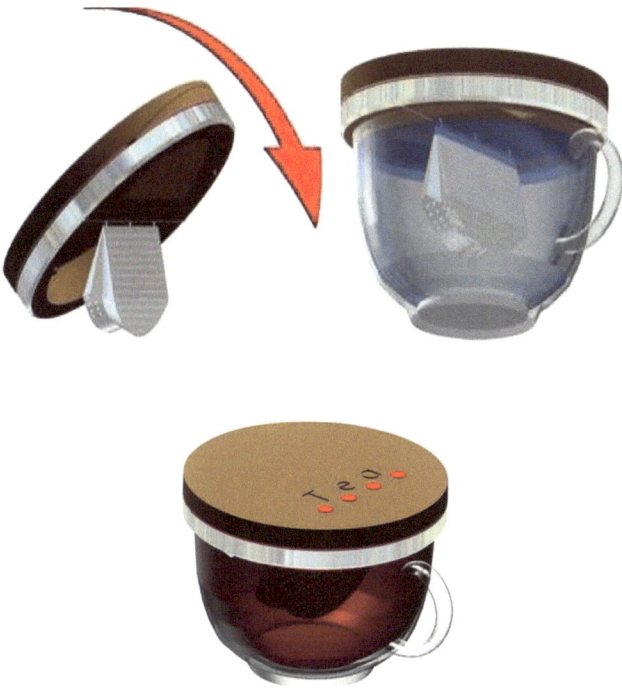

- This Package is Designed to Recycle Tea Bags and Maximize Flavor.

- This eco-friendly tea package has a separate vessel that holds the tea bag until it is ready to be used and helps maintain the original aroma and taste of the tea.

- This tea package is designed for the taste buds, it also has an eco-friendly element as the cup is able to hold the tea bag in a separate area from the actual tea, which allows us to reuse and recycle the tea bag more than once.

TEA PACKAGING INNOVATIONS

Manufacturer/Designer

Cedrik Ferrer
Republic of Tea

- Circular Tea Pods a new variation in tea bags.

- Each individual flavors contain its own illustration that has a consistent drawing style throughout the whole system.

- Redesigning the Republic of Tea by transforming the brand and packaging into a meaningful, modern, laid-back, and consistent design and creating an abstract version of their signature round tea bags as the logo; It creates a story to their company's identity and brand.

TEA PACKAGING INNOVATIONS

TEA PACKAGING INNOVATIONS

Manufacturer/Designer

Soon Mo Kang
Hanger Tea

- The tea bags are shaped like T-shirts with a hanger that hangs on the tea cup.

- Each tea type is highlighted in a different color for ease of tracking.

- Designer Soon Mo Kang invents the packaging of tea around mini clothes.

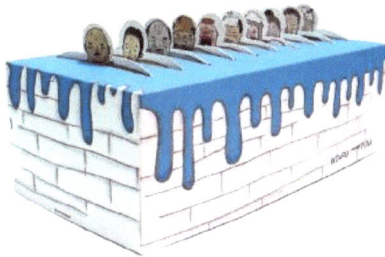

Manufacturer/Designer

WDARU design studio
Maum tea bags

- Maum tea bags fit most common cup sizes and come packaged inside beautiful tea boxes.

- It is a series of creative tea bags that look like little people with different personalities.

- These bags can be hung at cup of tea to make easy handling.

TEA PACKAGING INNOVATIONS

Manufacturer/Designer

Felix Reinki

- Tbag is designed in 'T' shape allows it to be hung from the sides of a mug, so that the tag does not drop into the tea.

- Elimination of use of thread saves material, cost and space also.

- Example of literal design applied to a tea bag.

TEA PACKAGING INNOVATIONS

Manufacturer/Designer

Elisabeth Soos

• tPod - Boats for tea bags.

• Tea bags hanging with a little paper boat that floats at the surface of the Parliament, an anchor made of tea.

• These tpod with a boat are packed in boat shaped carton.

Cigarettea

Don't smoke tea. Drink cigarettes

TEA PACKAGING INNOVATIONS

Manufacturer/Designer

Schnaider

- Cigarettea is cigarette-shaped tea bags for people who like to entertain in a subtle way.

- The "filter" will act as a floatation device.

- People will give glances of confusion as they drop a cigarette into boiling water.

- Thin tube bag shape is easier for transportation if anybody can't live without tea time.

- Don't smoke cigarettes ... drink 'em! Here's an idea so crazy it's brilliant: Cigarettea by Schnaider.

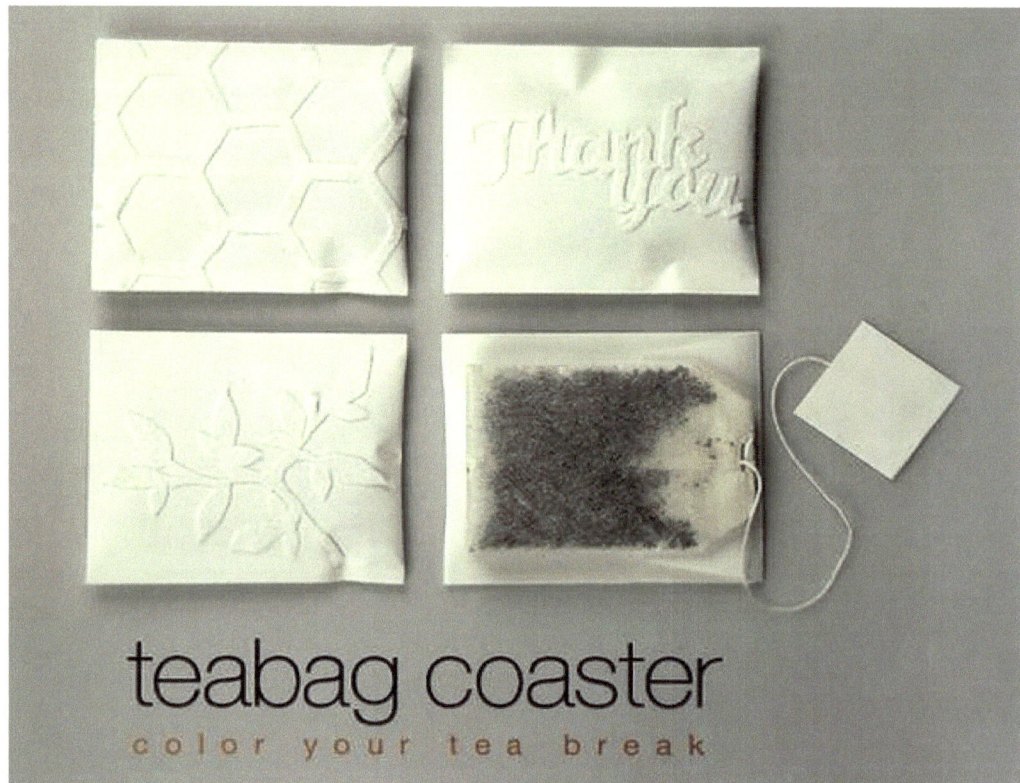

teabag coaster
color your tea break

Manufacturer/Designer

Yuree S. Lim & Jieun Yang

- Up-cycling tea bags.

- Teabag coasters create unique pieces from old tea bag.

- Simply remove the tea bag from cup, and put it on one of the coasters. Allow the remaining tea to soak into the teabag coasters and wait for the art to develop.

- These teabag coasters are a really unique and cool way to make use of used tea bag, and they don't involve putting them on our eyes to take away puffiness.

- These neat little contraptions will keep our table clean.

TEA PACKAGING INNOVATIONS

net weight 50 g   25 figured tea bags

GREEN
BERRY
TEA

Flavor combines the delicate sweet taste of our steamed kencha leaves with berries. Naturally rich in antioxidants.

100 % natural

GREEN
BERRY
TEA

TEA PACKAGING INNOVATIONS

net weight 50 g   **25** figured tea bags

GREEN
BERRY
TEA

100 % natural

GREEN
BERRY
TEA

Manufacturer/Designer

Natalia Ponomareva
(Russia)

- Each creative figured tea bag looks like a beautiful origami bird.

- These origami bird shaped tea bag are packed in small pouch with simple printed pattern and brand name.

- Each carton of tea contain 25 figured tea bag with simple graphics printed on it with cool color combination.

Manufacturer/Designer

ITCM
Unilever®

- PG Tips pyramid® tea bag are pyramid shaped tea bags which is an innovative design for tea bag packs.

- This was followed by the design and development of a patented, high-speed manufacturing process, including the ultra fast tea bag machine and the equipment for the novel 'caddy' carton format.

- The unique manufacturing process has been honored in the British Library book "Inventing the 20th Century" as one of the top 100 global inventions of the last 100 years.

TEA PACKAGING INNOVATIONS

- This use and throw bag is self efficient.
- Open the cup, the pod slide out dip and prepare the infusion and re pack the used pod into the cup.
- Hassle free and easy to use on the go, at work stations, during meetings.

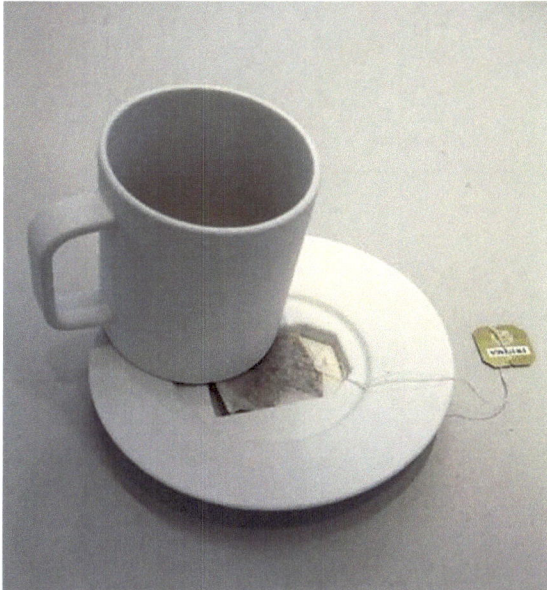

Manufacturer/Designer

Jonas Trampedach

• Now do not wait whole tea-bagging life for a place to bury tea bag.

• With the 'Tea bag Coffin', the drinker can tidily bury the bag under the cup and out of the way.

• This tea coffin has been designed by observing the behavior of tea drinkers and developed a solution to the bag dilemma that is as simple as it is ingenious.

TEA STICK
AND
STIRRER PACK INNOVATION

TEA PACKAGING INNOVATIONS

Manufacturer/Designer

Amcor Flexibles

• Tstix® is an innovative alternative to tea bags and pods. More than 1,100 micro holes create a clean and convenient tea experience – no spoons, no mess, no drips and no fuss.

• Consumers can now enjoy a cup of tea at any time, no matter where they are. Simply put the Tstix® in a cup of hot water and stir until the drink has the desired strength.

• The micro-perforated stickpack enables flavours control, is convenient and simple in its use and leaves no mess behind.

• The Tstix® concept is not limited to applications in the tea segment. This unique packaging solution can be applied in creating new packaging for coffee or fruit flavoured beverages.

• Amcor Flexibles license holder for the micro-perforated Tstix® package, is proud to announce that this new and inventive packaging solution has won The President's Award Bronze at the recent WorldStar Packaging Awards.

TEA PACKAGING INNOVATIONS

TEA PACKAGING INNOVATIONS

The Clever Quikstix Device

BC1   QUIKSTIX®

The handle section contains sugar (or whitener). First, tear off the end and tip the contents into your cup and add hot water.

The clever micro perforated section contains real coffee or tea that infuses when you stir it in the hot water!

Manufacturer/Designer

Dave Hopper

- On one side , the tube is perforated and holds coffee grounds or tea leaves that are infused when the stick is stirred in the liquid. The other side, with a tear-off end, holds fresh beverage.

- Quikstix in BC1 (coffee with one sugar) and BT1 (tea with one sugar) varieties was launched in 2007 in camping stores across Australia, where it has met with great acceptance.

- This segmented tube holds all the ingredients needed for a sweetened cup of coffee or tea. Tear, tip, stir, and sip are the only instructions needed to prepare a cup of coffee or tea with sugar using Quikstix.

Tea Stick/Stirrer

Manufacturer/Designer

Lee Yun Qin

• Tea stick seeks to re-define the form of a tea-bag, such that it can also act as a stirrer for the user, allowing the user to enjoy the tea through stirring it easily.

• The simplicity of the form also allows it to be hung from the sides of a mug, so that the tag does not drop into the tea.

- Cool tea bags designed for people who do not make their tea in a kitchen like gardeners, builders and campers etc.
- Tea stick is designed so that it can be used as a stirrer also.

CALENDAR TEA BAGS

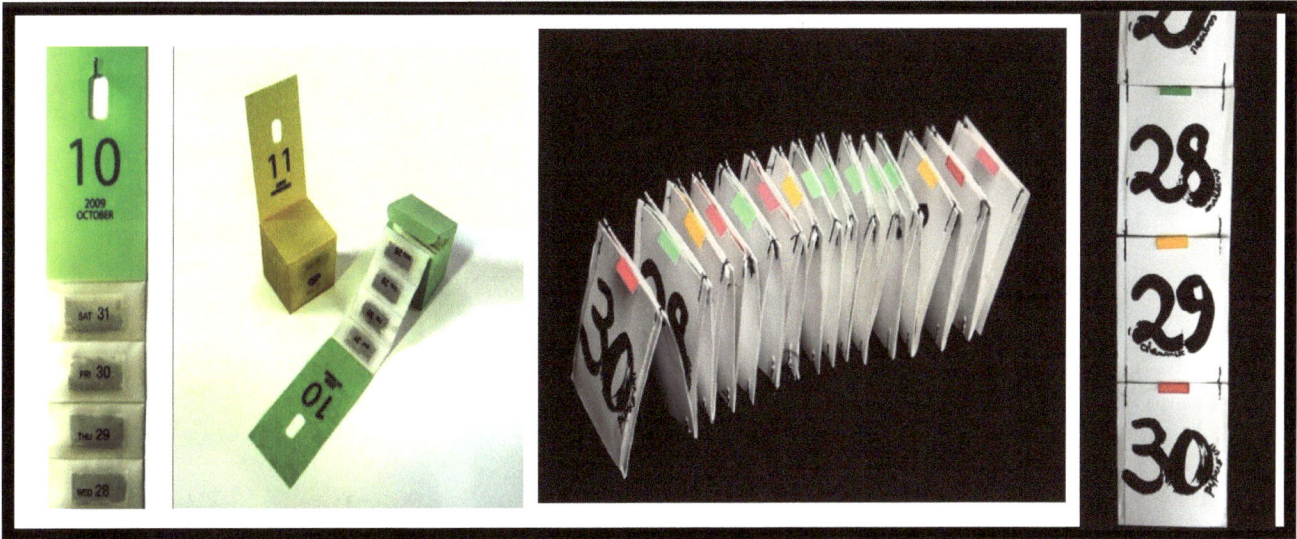

TEA PACKAGING INNOVATIONS

Manufacturer/Designer

Cho Hee Ha

- Calendar Teabags are basically a box of teabags with a variety of teas packaged as the dates of that month.

- A novel idea for tea buffs and those who appreciate niche products.

TEA PACKAGING INNOVATIONS

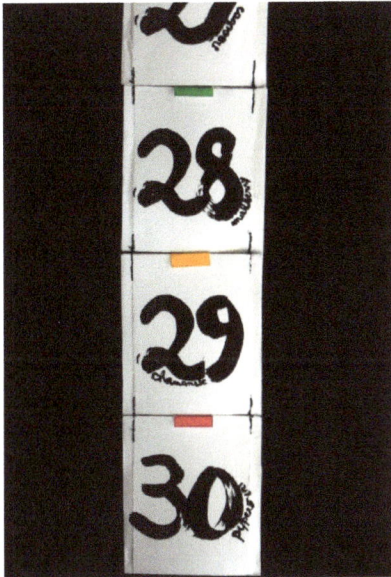

Manufacturer/Designer

Beatriz Barros

- Tea bags for 30 days of a month.

- Each tea bag is printed with dates of month so that consumers can pick their tea bag according to date of month.

- 30 DAYS TEA ADVENT Contest Project Packaging for a tea advent.

OPEN TEA BAGS

TEA PACKAGING INNOVATIONS

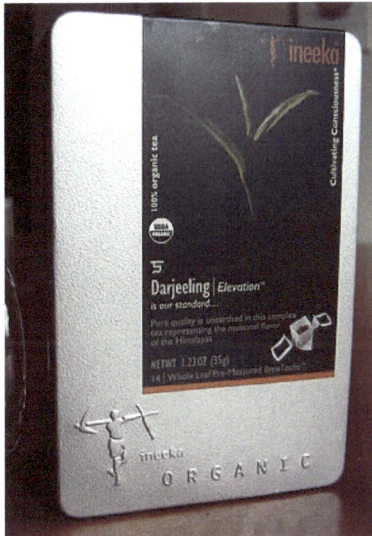

Manufacturer/Designer

Ineeka Darjeeling Tea

• The bags have paper arms on the sides that fold out in order to make them into a single-use filter.

• The tea inside is effective loose because before placing the tea in the cup, tear off the top and the let the leaves float around in the open bag while it steeps.

• By opening the top we can also pour the hot water directly onto the leaves without them scattering, which helps

Manufacturer/Designer

Cafusa™

- The patented Pour-thru™ bag design allows consumers to see the product as they brew it, which is highly desirable for specialist herbal and long-leaf teas.

- By opening the top, we are able to pour the hot water directly onto the leaves without them scattering, which helps to make a fine beverage.

- A tea bag for customers who are looking for new and innovative coffee and tea packaging.

TEA PACKAGING INNOVATIONS

INNOVATIVE TEA PACKAGING CARTON

TEA PACKAGING INNOVATIONS

- There can be shooting stars, rock stars, film stars, star fish, star wars, mega stars, and now there's Teastar™.

- There are 10 organic loose leaf teabags, wrapped together to show all human emotions to form a star shaped box.

- As per emotion the tea bags can be opened!

- Once opened, the box is like an old fashioned paper yap toy. Peer inside and choose favourite brew.

- Each Teastar contains 2 x Happiness, 2 x Sleepy, 2 x Inspiration, 2 x Love and 2 x Friendship loose leaf teabags. Choose your cup of tea!

TEA PACKAGING INNOVATIONS

TEA PACKAGING INNOVATIONS

Manufacturer/Designer

Polina Sapershteyn.

• New "box" holds a roll of tea bags which may be replaced over and over again, eliminating superfluous packaging.

• Structural design does not only serve to distinguish product on the store shelf, but allows for an environmentally-friendly use.

• Biodegradable rolls of tea bags need only to be purchased, while the dispenser is kept for repeat use.

• The natural texture patterns represent the brewing of the tea and the delicacy of taste, or the "winged B" logo alludes to the elevated tea experience. "

TEA PACKAGING INNOVATIONS

TEA PACKAGING INNOVATIONS

Manufacturer/Designer

Maria Milagros Rodriguez Bouroncle

- At the first look this black box looks like probably a wine packaging.

- The minimalist black box opens to reveal an explosion of colourful sachets, That contain tea bags.

- Each one a piece of paper folding art in itself, and able to be recycled.

- Each colored sachets carries single tea bag.

- Unexpected and surprising contrasts between outer carton color and after opening colorful sachets seem so fundamental to the opening of packages.

- Packaging Determined to combine eco-efficiency with simple yet striking aesthetics.

TEA PACKAGING INNOVATIONS

TEA PACKAGING INNOVATIONS

- The idea behind hexagon box was to connect tea and fang shui energies. So each side of hexagon represents 5 energies (water, fire, earth, metal and wood) and sixth is the Dina that joins all the energies to create a tea with the exquisite taste.

- Package of each tea taste uses the color that represents one of the fang shui energies.

- Simple shapes, black background, pattern are meant to achieve exclusive and Asian feeling.

- Hexagon box can be split into triangles inside to fit pyramid tea bags. Hexagon form is easy to stack on a shelf and can be combined to form different patterns for presentation.

TEA PACKAGING INNOVATIONS

Manufacturer/Designer

Rationale

- Japanese art of origami is incorporated in this design, having the lid fold together.

- When closed, it looks like a small bunch of tea leaves.

- The design is clean and simple, just like the tea it holds.

- Simply 'T' is a Japanese tea company specializing in green teas designed three tea packages within a series. This line includes jasmine, matcha, and hojicha.

TEA PACKAGING INNOVATIONS

teatime Organic Loose Leaf Tea Jars

OOLONG

GREEN

RASPBERRY

4 Pack

4 4oz. Jars NET WT 16oz.

TEA PACKAGING INNOVATIONS

Manufacturer/Designer

Kayla Bishop

• Pattern is designed on the box to be organic and light to be easy on the eyes but kept green to grab the attention of the customer.

• The package includes a variety pack of four loose leaf teas in reusable jars.

• The tea would be sold where customers could bring their jars back to be refilled.

TEA PACKAGING INNOVATIONS

- The box that holds the leafy beverage pouches is multifarious. It comprises more generally of a square inset carton but each of the teabags inside of it are wrapped individually into different geometric shapes.

- Chee Teachee Tea packaging contains the opportunity to sit back, relax and share a delicious drink with a friend, and offers its sippers a little game to play.

- Dho Yee Chung's Chee Teachee Tea packaging is a Chinese tan gram puzzle that will keep us amused from first sip to the last.

TEA PACKAGING INNOVATIONS

TEA PACKAGING INNOVATIONS

Manufacturer/Designer

Abstract Group

- Using a curved Truncard folding box design and a large colorful palette adoring a white box, Flora Tea's packaging is now as eye-catching.

- The physical design works very well; the floral design for the box resembles a closed flower and when the box is opened it looks like an open flower.

- The Trucard folding box board matched the needs in providing a good print surface and bi-directional folding.

TEA PACKAGING INNOVATIONS

Manufacturer/Designer

Belinda Shih

- The design is focused on the concept of afternoon tea - a time to unwind, re-energize, and relax after a long day of work.

- The package is therefore minimalist, calm in color and simple in shape, with a hint of the soothing elegance of steam swirls and tea drops.

- The container is reusable to encourage customers to buy replacement tea bags in the future for resource and money conservation.

- This is a packaging design concept for a hypothetical tea company.

Manufacturer/Designer

Winni Hsiao

- The packaging design is for Kandi Organic tea. The design concept is the idea to play with similar pronunciation from the word "candy".

- Vibrant color and the twist top to reflect the slight sweetness in the herbal tea. In each box there are 15 packs of tea bag wrapped individually.

- The individual tea packaging can be reused as a little tray when the tea bag is soaked in the water. Each color of the packaging also represents the meaning of the three flavors.

TEA PACKAGING INNOVATIONS

Infographics

Manufacturer/Designer

Caren Sutton

• Each side of the tea box contains tea specific infographics regarding both general facts about tea as well as specific information about chai spice black tea.

• The bottom of the carton has a die-cut for easy pull out of tea bags.

• The color palette and design elements are inspired by Indian henna designs as well spices to create an informative yet pleasing design, both visually and sensually.

Manufacturer/Designer

Andrew G. Herbert

- Three component pack.

- 1st is Carton having a sliding flap at bottom to draw out tea bags.

- 2nd is tea mug.

- And 3rd one is tin canister in which both of carton and mug are to be packed.

- Halcyon Tea Packaging has received accolades and recognition from several competitions and showcases, Including, Pantone® COLOR in Design Awards 2012.

- Point of debate maybe, is it really an innovative pack or wastage of packaging?

TEA PACKAGING INNOVATIONS

Manufacturer/Designer

Nicholas Wright (Fort Hays State University)

• The package design resembles butterfly display cases to present the product.

• A butterfly's beauty is hidden until it transforms and emerges from it's cocoon.

• The brand identity is inspired by the standard scientific naming convention by combining two Latin words to create the meaning "hidden beauty".

• Much like the butterfly, tea is not visually attractive, but when brewed it changes into a thing of beauty with unique aromas and flavors

TEA PACKAGING INNOVATIONS

TEA PACKAGING INNOVATIONS

Manufacturer/Designer

Sarah Déry

• Design Inspired from Rubik's Cube.

• The entire package is made from a single sheet of paper, folded and cut to allow the user to detach each cube as needed.

• Each cube has cubical shaped tea bags, which are ready to be brewed.

• A concepts which combine visual impact and emotional response.

• It is an environmentally friendly tea package, containing 27 cubes of compressed tea.

• A really nice design which allows to pack together different blends of tea in individual packaging.

TEA PACKAGING INNOVATIONS

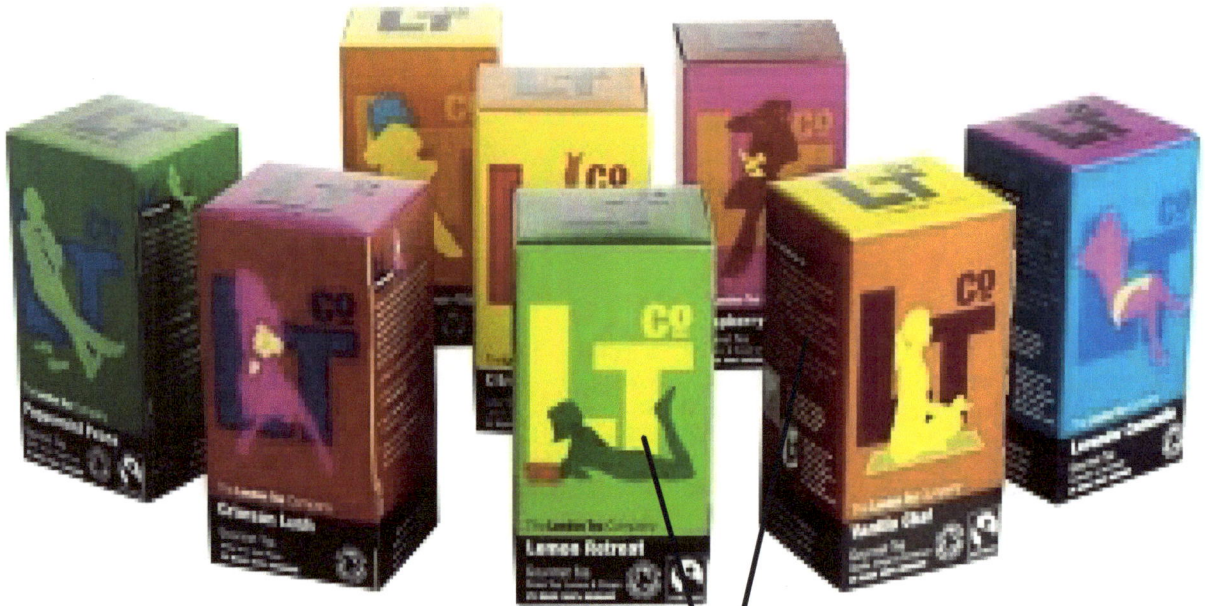

Images

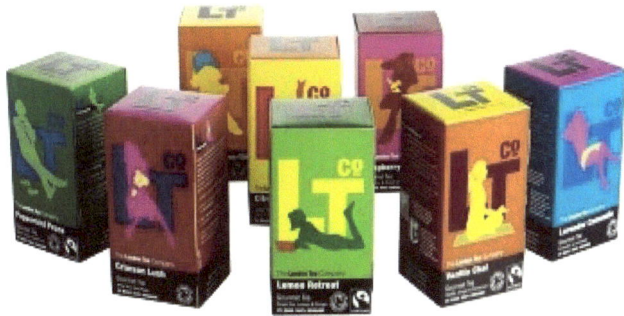

- Eco-friendly tea packaged using no glue, cellophane, wrapping or sachets.

- The boxes are made of recycled paper and printed with bright eye-catching colours.

- The images feel a little rushed, and possibly a little pigeon holing.

- These images denote the mood and type of person who should drink each type of tea?

- The style of design feels generic, not aimed at a specific audience.

TEA PACKAGING INNOVATIONS

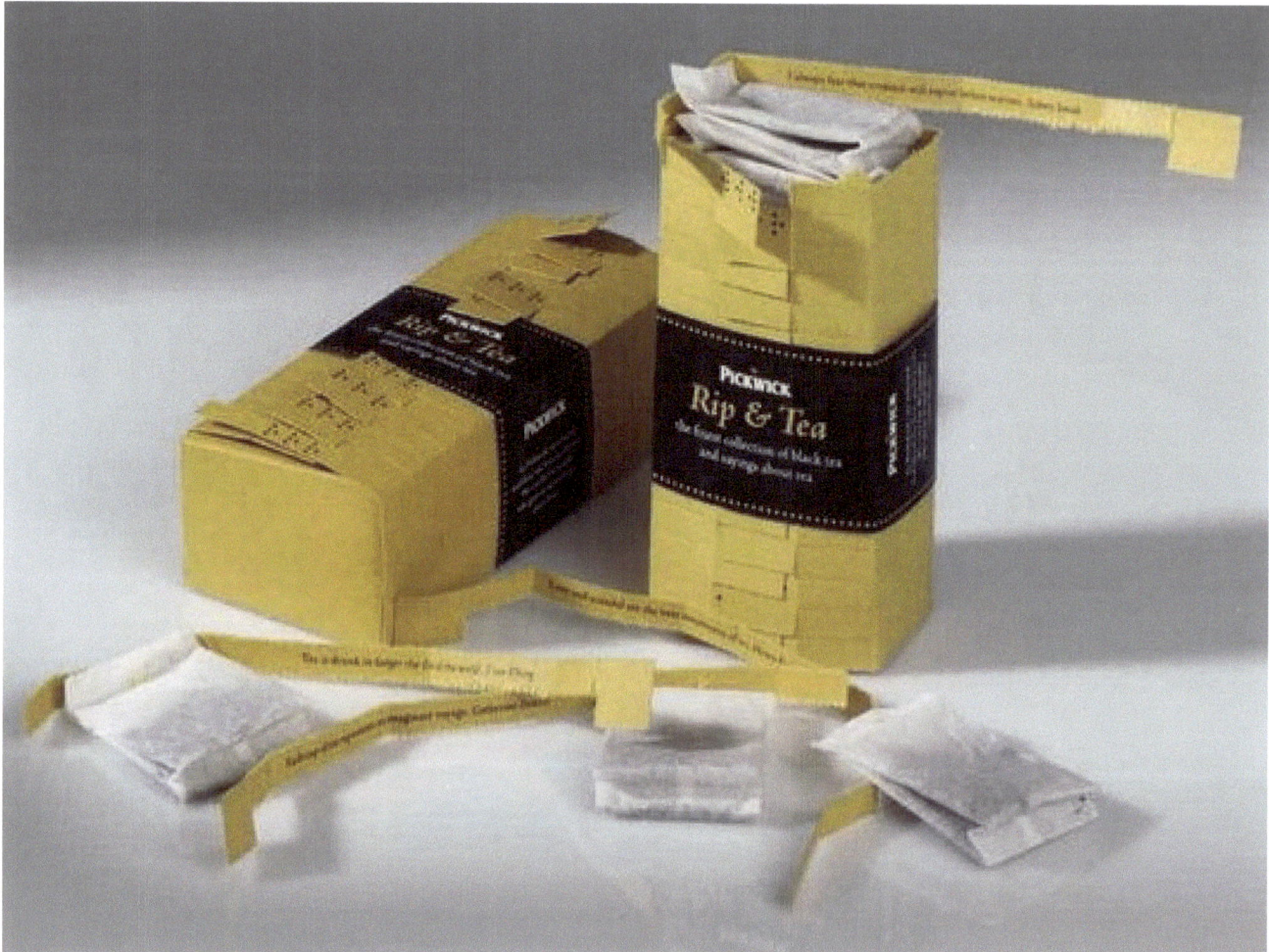

TEA PACKAGING INNOVATIONS

Manufacturer/Designer

Veronika Pethő

- This innovative carton is made for the dip tea category.
- The whole carton is perforated and each perforation has one dip tea sachet attached to it.
- The front of carton has the pull tab for each perforation, to separated from the carton.
- This perforated portion then can be directly used as the dip sachet, as shown in the picture.
- A very unique method of packaging, with zero packaging waste at the end.
- This can be a tamper evident and counterfeit packaging as well.

TEA PACKAGING INNOVATIONS

Manufacturer/Designer

Samantha Hartman

- Unique colors, font, and pack style combinitation make it classic packaging design.

- Each carton has tea pouches of respective color, font and flavours.

- There is die-cut at the bottom of tea pack which provides easy pull outlet for tea bag pouches and give it attractive and different look on the selves.

- Five different flavors in modern representation of a retro style through the font and colors.

TEA PACKAGING INNOVATIONS

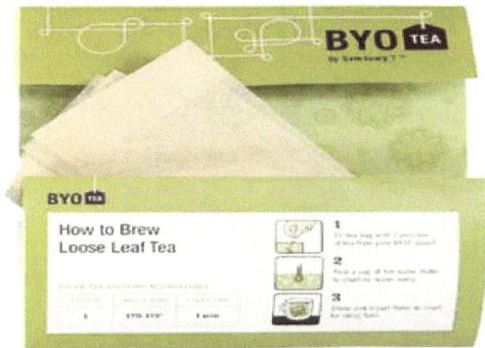

Manufacturer/Designer

BYOT' from the Sanctuary T Shop

- It is Ziploc bag full which is very sophisticated.

- This pack is a best travel companion, with carrying 10 tea bags each.

- It brings favorite flavors with us while out to lunch, on the plane, to a friend's, or wherever we'll get our leafy caffeine craving.

- These teas come in biodegradable pouches so one can return the earth's favor.

- The package is designed for cold brewed tea bags, which are immersed in cold water .

- It is long rectangular shaped tea bags, allowing consumers easily drop into or take out from PET water bottles.

- The idea is originated from the "ice flake" and applied its beautiful shape as well.

- It is designed to evoke the feeling of coldness.

- The latch design makes the structure strong, beautiful, and eco-friendly.

TEA PACKAGING INNOVATIONS

TEA PACKAGING INNOVATIONS

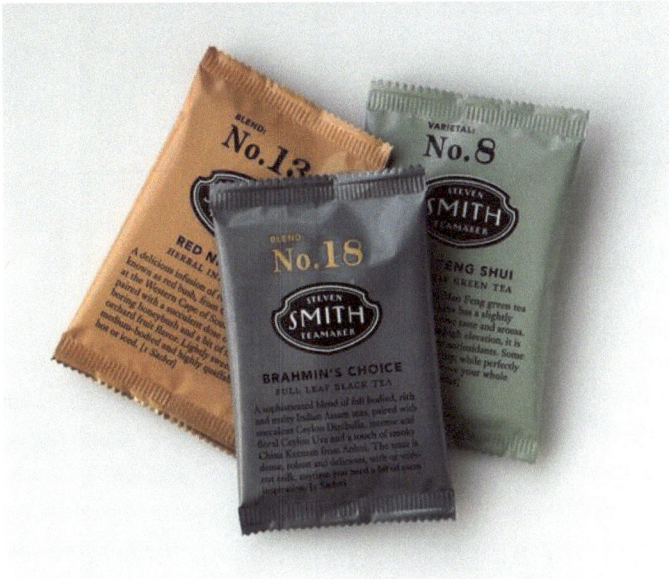

Manufacturer/Designer

Sandstrom Partners

- There is an innovative use of button and string for closure on opening flap.

- The pack contain pouches of tea for different flavours. Colored label on carton tells about the flavor of tea packed in it.

- Each pouch is packed with tea bags.

- Very elegant packaging for the new, eponymous tea brand.

TEA PACKAGING INNOVATIONS

Manufacturer/Designer

Ceremonie

- Decadent and opulent packaging, using Colour and texture to create a feeling of refinement.

- The decoration is quite intricate, but is paired with simple type which allows it to stand out.

- Opening flap of carton is much like cake box.

- It seems to go just far enough to create a feeling of a luxury product.

- Assorted, right about color dots

TEA PACKAGING INNOVATIONS

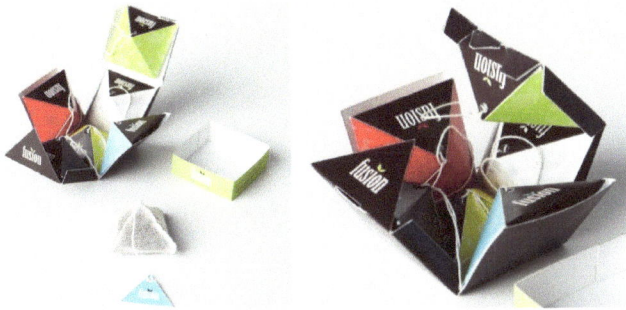

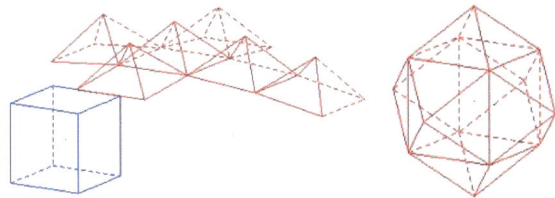

- Annabelle Soucy's polyhedral 6-pack for tea.
- Soucy's cube-shaped "Fusion" tea package dissects into 6 space-filling pyramids.
- Each section containing a pyramid-shaped tea bag.
- A simple exterior and complex interior, make this a structural "surprise package".
- A perfect pack to engage the customer and gives a brilliant SMOT*

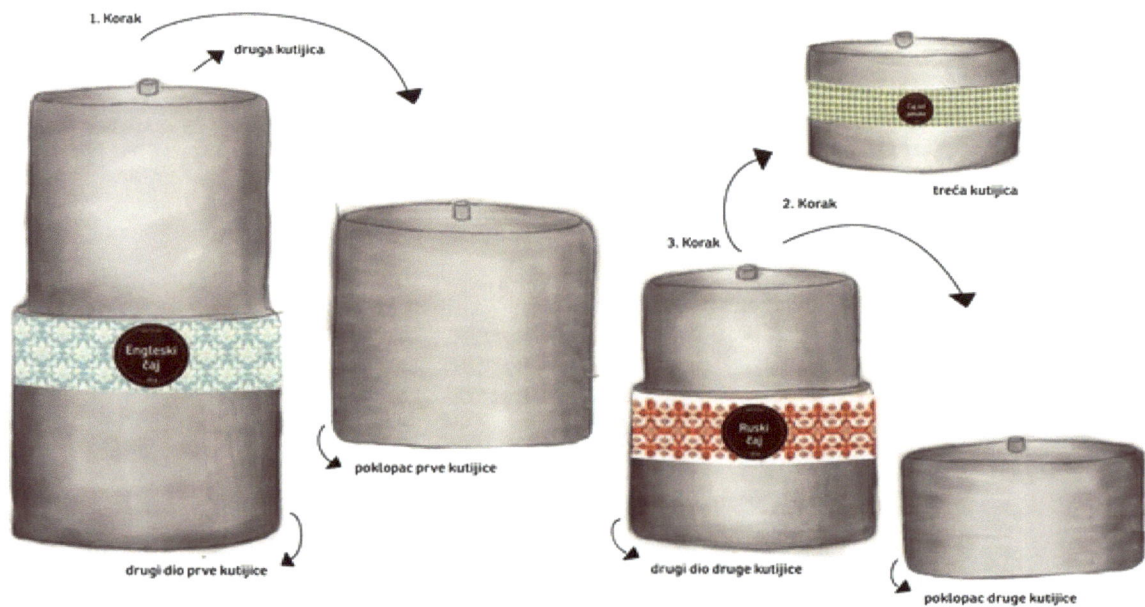

1. Korak

druga kutijica

Engleski čaj

poklopac prve kutijice

drugi dio prve kutijice

treća kutijica

2. Korak

3. Korak

Ruski čaj

drugi dio druge kutijice

poklopac druge kutijice

TEA PACKAGING INNOVATIONS

Manufacturer/Designer

Russian tea

• This work is inspired by wooden puppet babushka (series of 2 part puppets, small one goes into bigger one, there can be a lot of them in one) which comes from Russia as well as one of the most famous tea's - russian tea.

• Packaging is adoptable to every manufacturer regarding the combination of tea's they want to sell.

• Point is that the packaging is universal and special at the same time, it is natural and yet has a modern design.

INNOVATIVE
TIN TEA PACKAGING

Unique Tea quote

Manufacturer/Designer

Manic Design

- Vibrant collection of tins tea packaging with attractive color labels.

- All tins pack are with their own (labels) matching sticker seals and unique tea quotes.

- Each label features its own unique brewed leaves infused on a block-Colour background.

TEA PACKAGING INNOVATIONS

Taiwanese regional pattern

TEA PACKAGING INNOVATIONS

Manufacturer/Designer

Taiwanese branding and design studio Victor Design

- Beautiful identity packaging design
- The pattern design of the tea boxes is inspired by the Taiwanese regional culture and the beautiful landscape.
- These regional inspired patterns or neat and fine printing are specialty of this tin pack.
- These tin packs are packed in an another pack also.

Stacking
Combo
pack

Stacking
tea tins

TEA PACKAGING INNOVATIONS

- There are two packs of tin.

- Left pack in picture is biscuit tin combo pack made up by stacking tins for those people who love to eat biscuit with tea.

- Right pack in picture is stacking tea tins.

- Packaging Innovations Birmingham 2012

- Tinplate Products attracted plenty of attention to this tea.

TEA PACKAGING INNOVATIONS

Manufacturer/Designer

O-zone
Green tea

- Alignment of 'green tea', decorated with simple evocative designs.

- Surface picture behind green giving a sense of product itself that is tea gardens.

- And bright green graphics on dull background individualize the pack with an unique contrast.

- Fresh bright green itself giving a sense of green tea.

Effective Printing Designs For Tea pack

TEA PACKAGING INNOVATIONS

Manufacturer/Designer

Chris Yoon

- Design inspired by dry British humor, and created as set of characters from an existing line of teas from Twinings.

- The concept was to "get into the head" of each of the characters, and so the cylindrical container lifts off.

- It gives us a quip printed on each tea bag.

- These tea bags are circular pods which is an innovation in itself.

- Bold text on the pack is adding value to this pack.

TEA PACKAGING INNOVATIONS

TEA PACKAGING INNOVATIONS

Manufacturer/Designer

BYOT' from the Sanctuary T Shop

- Ziploc bag full which is very sophisticated.
- This pack is a best travel companion, with carrying 10 tea bags each.
- It brings favorite flavors with us while out to lunch, on the plane, to a friend's, or wherever we'll get our leafy caffeine craving.
- These teas come in biodegradable pouches so one can return the earth's favor.

TEA PACKAGING INNOVATIONS

TEA PACKAGING INNOVATIONS

- The colours are very strong which helps the overall design to become more eye-catching.
- It uses plants to elude to 'organic'
- But contrasts this with the heavy type and bold design.
- Heavy and bold type faces are playing the main role for branding and making it a unique product on selves.
- This design is quirky.

TEA PACKAGING INNOVATIONS

Aluminium

Cardboard

Manufacturer/Designer

Caribou Tea Packaging

- The old tea chest gets a charming reference in Caribou Tea packaging. Despite the boxes being made of either aluminum or cardboard, they bear the detailed dressing of wooden structures, this line of products does communicate nostalgia and an earthiness that entices infusion enthusiasts.

- The image of lovely yellow wood grain covers every face of the miniscule crates and the graphic effect seems to reveal the wood's natural texture behind the stamped logo.

- An illusion of applied paper labels to Caribou Tea packaging provided a visual opportunity to include more information on the beverage variety inside, enhanced with endearing illustrations.

TEA PACKAGING INNOVATIONS

TEA PACKAGING INNOVATIONS

Manufacturer/Designer

Natsume Matcha

- The packaging reflects the calming and slow growing process of the tea.

- The coloring is soft and the packaging very simple and has a very soft and calming feeling.

- Fine type face and cool color combination make this pack more attractive and cool.

- A top quality Japanese tea used it in traditional tea ceremony.

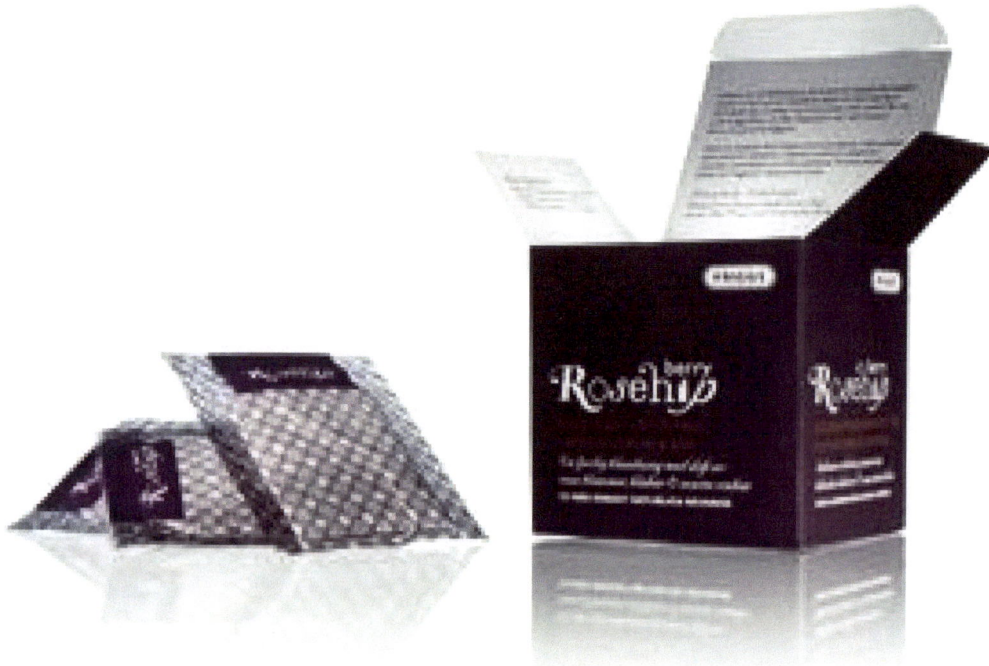

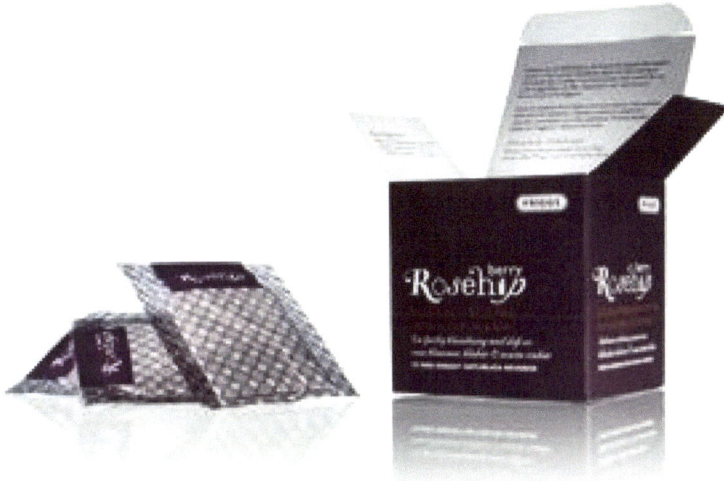

Manufacturer/Designer

Rosehip

- This design uses very plain, generic packaging.

- The type is mixed and creates a sense of fun.

- While the dark coloured background allows the type to stand out well.

- Dark color also make the pack royal as well as simple.

- Simple packaging with a pretty design.

TEA PACKAGING INNOVATIONS

Manufacturer/Designer

Peyton and Byrne

- These all packaging has a very traditional and British design.

- Packaging uses bright, defined coloured and has a very minimalist approach.

- The strength of their design is the simplicity and uncomplicated approach.

- There is a sense of class within this design.

TEA PACKAGING INNOVATIONS

Manufacturer/Designer

Ku Kai

- A very simple approach to the design, using illustration and bold colours to differentiate between flavours.

- 'KU KAI' type faces used in in pack are unique specially for letter 'U', which also resembles a cup graphic design is illustrated which is adding value to branding of tea.

- White background for type faces is adding more readability and leasability.

Mallard

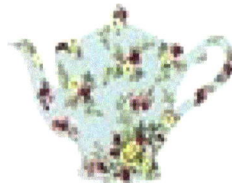

Mallard

Mallard

Mallard

Mallard

Mallard

Manufacturer/Designer

Mallard Tea Rooms

- This mix and match style of design for Mallard Tea Rooms is inspired by the eclectic style of the tea rooms themselves.

- Unmistakably British and eccentric the personality is strong, but simple and clean.

- Number of patterns are used for different flavored tea.

- Appealing to a small audience, those who appreciate proper English tea service and those who would feel that proper tea is better tea.

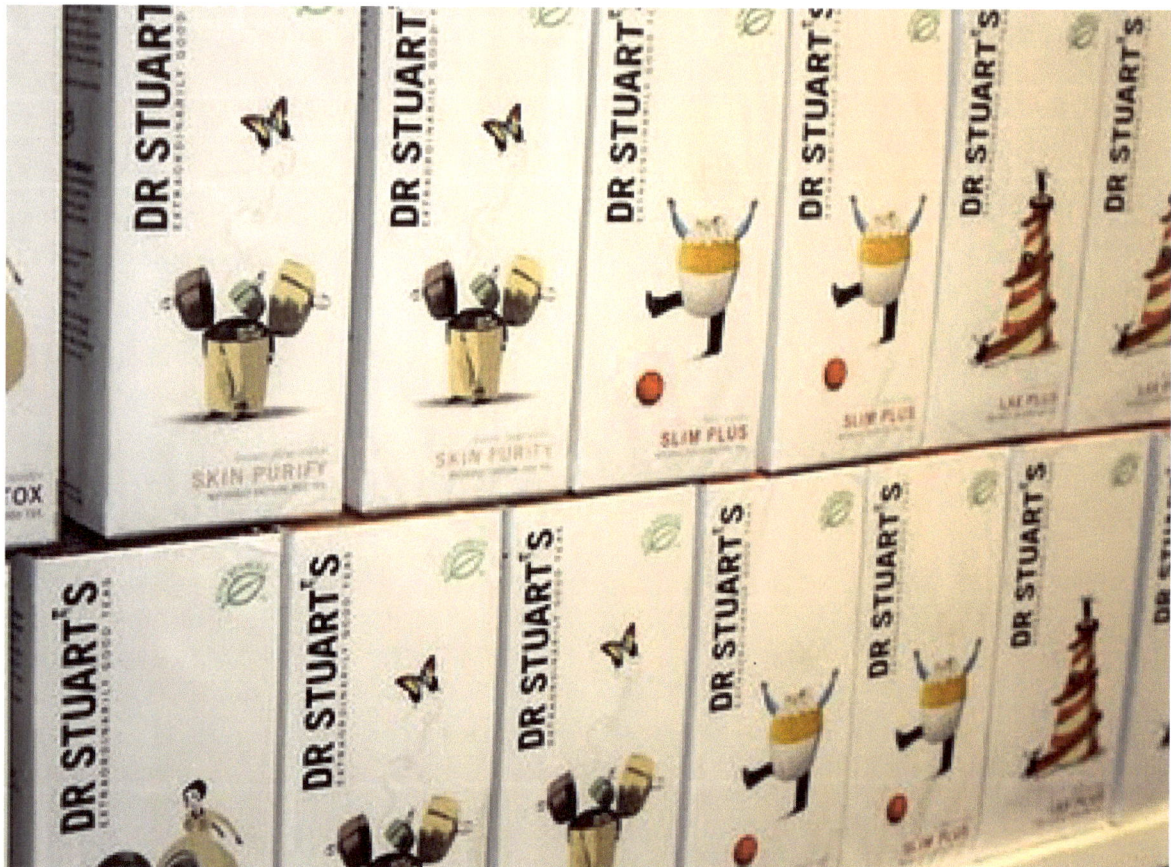

TEA PACKAGING INNOVATIONS

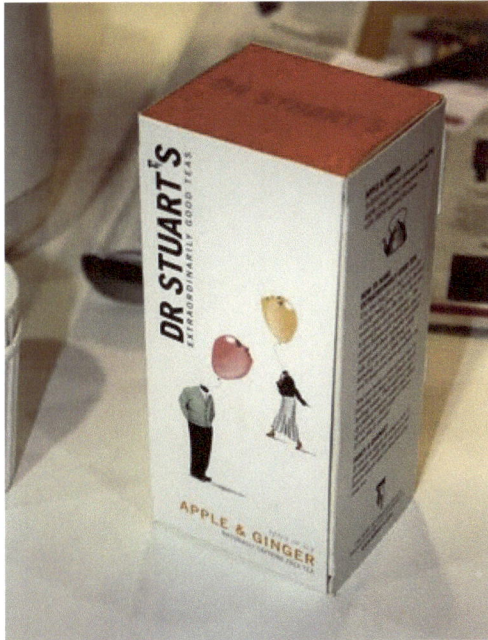

Manufacturer/Designer

Dr. Stuarts

- The quirky illustrations and clinical packaging is eye-catching and very strong.

- The simplicity of pack and illustrations are making this tea product clinical.

- The graphic design is itself projecting about its customers that who should buy this product.

- The overall personality is fun and friendly and very engaging.

- The use of Colour is also very good, drawing attention to the illustrations first and the branding second.

TEA PACKAGING INNOVATIONS

TEA PACKAGING INNOVATIONS

Manufacturer/Designer

Andrews & Dunham Tea

• These tea canisters are vibrant and bold.

• The bold colours are eye-catching and enhance the overall sense of fun within the product.

• Different design and color combination are used to tell about different flavours of tea.

• Each design is completely different which allows for strong and varied design styles.

TEA PACKAGING INNOVATIONS

ORGANIC WHOLE-LEAF TEA

TEA·hugger

JADE DEW (GYOKURO)

EARTH FRIENDLY TEAS

GREEN TEA

Forty Satchels

NET WT. 3 OZ (88g)

TEA PACKAGING INNOVATIONS

Manufacturer/Designer

Tea Hugger

- This packaging is a combination of modern and traditional ?

- The shapes, pattern and colours well very traditional.

- But the type is modern and has a hand written quality to it.

- Overall the design is interesting; it is engaging and has something quite quirky about it.

TEA PACKAGING INNOVATIONS

Manufacturer/Designer

Pavla Chuykina
Teatul Tea

- Adorable packaging design for packaging of loose leaf tea.

- Different colors are used according to different flavors to add more beauty.

- Specially different typefaces are used for Teatul for the purpose of branding.

- Cap is bigger one which goes in to smaller pack (Telescopic) is totally different concept of packaging.

TEA PACKAGING INNOVATIONS

TEA PACKAGING INNOVATIONS

Manufacturer/Designer

Sara Strand

- Same as the product, design is also illustrated according to child mentality to attract them.

- Use of vibrant colors and combination of cartoon cups and cartoon graphics are used to appeal children.

- Children's Tea comes in three different flavours, Sunny Orange, Red Berries and Wild Berries.

- Children's Tea is the perfect choice for the young ones. All ingredients are organic and naturally caffein-free.

TEA GIFT PACKS

Manufacturer/Designer

Chris Eshnaur

- Hand crafted wooden box made completely from scratch

- Pop-up ship created specifically for this packaging.

- Hand-done wood burned logo on box exterior.

- Wooden box is packed with tea bags.

- Paper crafted ship is attached with the thread of tea bags

- Overall design and tea bags make it a complete pack as the name suggest Sea Tea.

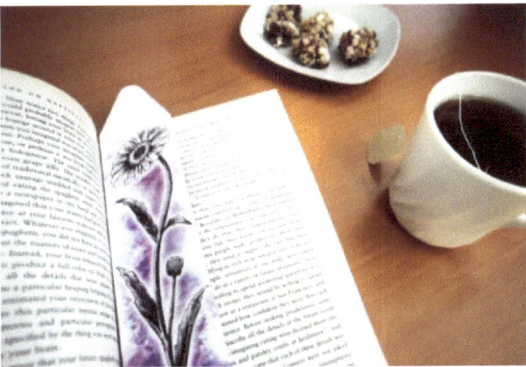

Manufacturer/Designer

Alexander Chin

- It is best gift for tea drinkers who read while brewing, the vertical orientation of the stick makes it a very convenient book mark."

- When the tea drinker removes the tea bag from the stick, a prompt followed by a small flower is revealed.

- The back of the stick corresponds by growing taller the farther a person needs to reach their next tea bag.

- While not only protecting the tea bags, the outer stick mimics a nostalgic white picket fence which displays the ingredients proudly on the back.

- The flowers are hand illustrated in attempts to create packaging too beautiful to throw away.

TEA PACKAGING INNOVATIONS

Origami tea bags

Manufacturer/Designer

YIU Studio.

- Dovely Tea is part of a line of handmade gift goods designed to be a unique and memorable.

- The package is hand made, using letterpress, silkscreen and stamping techniques.

- Then hand assembled including the origami tea bags i.e. origami bird structure is given to the tea bag.

- The color combination of bright color with black giving the pack a royal look.

- So, this is a kind of perfect gift for tea lovers.

TEA PACKAGING INNOVATIONS

Manufacturer/Designer

Proud Design

- Three Tea container packs packed in single sleeve.

- The tea leafs are placed inside a tin container.

- Each container also slide into colorful sleeves.

- Color of each sleeve reflects flavours of tea.

- Each sleeve is given a new name and image that reflects the individual characteristic of each tea blend.

- These can be the perfect gift with a personal touch.

TEA PACKAGING INNOVATIONS

Manufacturer/Designer

Jasen Melnick

- The package uses stylistic forms and patterns.
- Each pattern reflects central and eastern Asia to add cultural value to the tea.
- Cartons are display cartons which reflects the packed product or loose leaf.
- A die-cut at the top flap making easy handling of carton.
- Shape of carton is unique and used triangular shape to differentiate the product on selves.
- Packaging system designed to brand loose-leaf tea as a luxury product.

TEA PACKAGING INNOVATIONS

MANGO ROOIBOS TEA

RASP BERRY WHITE TEA

STRAW BERRY GREEN TEA

TEA PACKAGING INNOVATIONS

Manufacturer/Designer

Ralu Ciubotaru

- A beautiful Handmade cardboard tea boxes for gift.

- And graphics drawn on box look like hand painted or drawn with chalk.

- Use of different color combination to reflect the flavours of tea.

Manufacturer/Designer

Lun Yau

- The term 'food speaks for itself' approach lead a literal and visual approach.

- The packaging is inspired by chalkboards, which give an organic and personal feel.

- The packaging for tea and coffee used the same screen printed speech bubble package with the Delilah logo, and Coloured chalks are supplied within the hamper so that the packaging can be personalized as a gift.

- The gift hamper packaging show speech bubbles that can be customized by customer this adds a personalized gift element to their purchases.

TEA PACKAGING INNOVATIONS

Manufacturer/Designer

Emeyu Tea

- This is a tea 'library'.

- The bright Colour is engaging and the style of packaging presents a modern but informed personality.

- The idea of a 'library' seems a little excessive, but this may appeal to tea connoisseurs or those interested in broadening their horizons.

- The design feels very in tune with the origins of tea, the style and importance of ceremony.

- It feels like more of a gift rather than something one would buy from themselves.

TEA BAG INFUSERS

Tea Bag Tea Infuser

TEA PACKAGING INNOVATIONS

- Tea Bag Tea Infusers

- The Tea Bag Tea Infuser is the Waste-Free Way to Prepare our Drink.

- The best teas truly come in loose-leaf form, and the Tea Bag Tea Infuser helps to promote good quality brew with the perk of waste reduction.

- The stainless steel or plastic strainer in different shapes designed so that, it can be used over and over again and conveniently cleaned in the dishwasher.

TEA PACKAGING INNOVATIONS

TEA PACKAGING INNOVATIONS

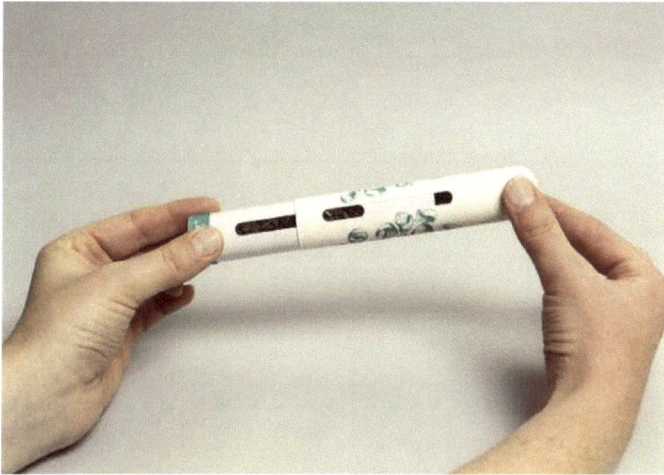

Manufacturer/Designer

Designer Lillian Cutts

- Lillian Cutts proposed two very different receptacles to encase the exquisite contents.

- The first is a delightful white tin decorated with rosy red illustrations. It is just wide enough to accommodate a line of teabags, labeled with tags of the same charming motifs.

- The sample tubes are quite unique and appear to be the perfect vessels for small quantities of the dried beverage.

- Twig & Leaf Tea packaging is endearing and artistic, and fit to impress consumers with five fine-tuned senses.

TEA PACKAGING INNOVATIONS

TEA PACKAGING INNOVATIONS

Manufacturer/Designer

Candy Cane

- The Candy Cane Tea Infuser

- This Candy Cane Tea Infuser would look adorable packaged in cellophane with a holiday mug and some peppermint tea, making it a thoughtful.

- Candy can packed so that tea leaf remain fresh in candy can by covering the holes of candy can completely.

- Feature: - Stainless steel 2-piece tea infuser, Dimensions: 1.8" x 4" x .6"

TEA PACKAGING INNOVATIONS

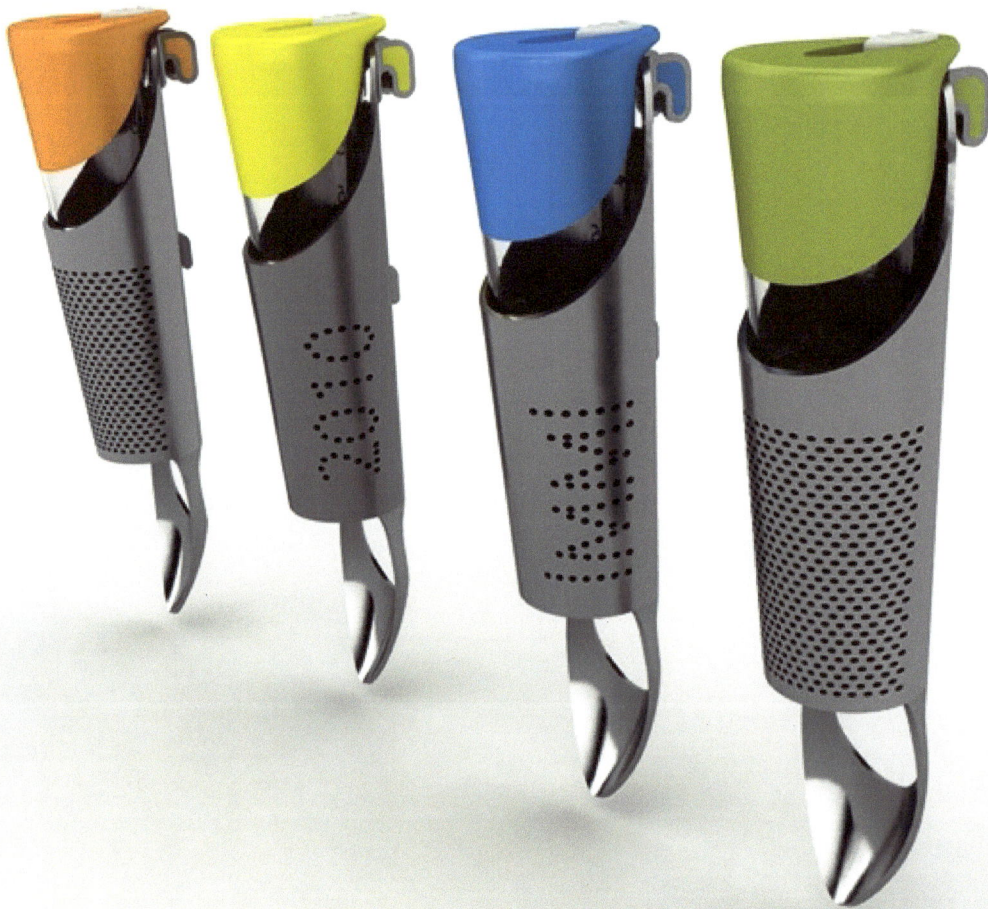

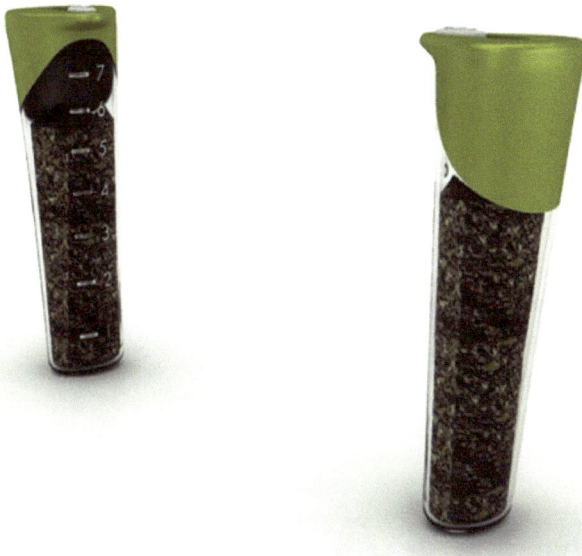

Manufacturer/Designer

Dr. Hakan Gürsu for Designnobis

- The infuser is made out of a unibody piece which is first cut from chrome stainless sheet metal and then bent like spoon.

- Therefore it does not need an extra process for assembly, so this infuser is also a stirrer itself.

- For the transparent container, polypropylene (PP) is used, while again PP is used for the remaining plastic parts in infuser.

- This PP container is shaped to fit in the body of infuser and can be re filled or reused.

- All plastic parts are produced with injection moulding.

TEA PACKAGING INNOVATIONS

COLD TEA PACKS

TEA PACKAGING INNOVATIONS

Manufacturer/Designer

®evolution

- Ready to drink white tea packaged in plastic bottles.
- The designs are coloured for differentiation.
- This carries through to the photographs denoting flavours.
- Overall the designs are very low key, they have little personality and use a minimalist approach.

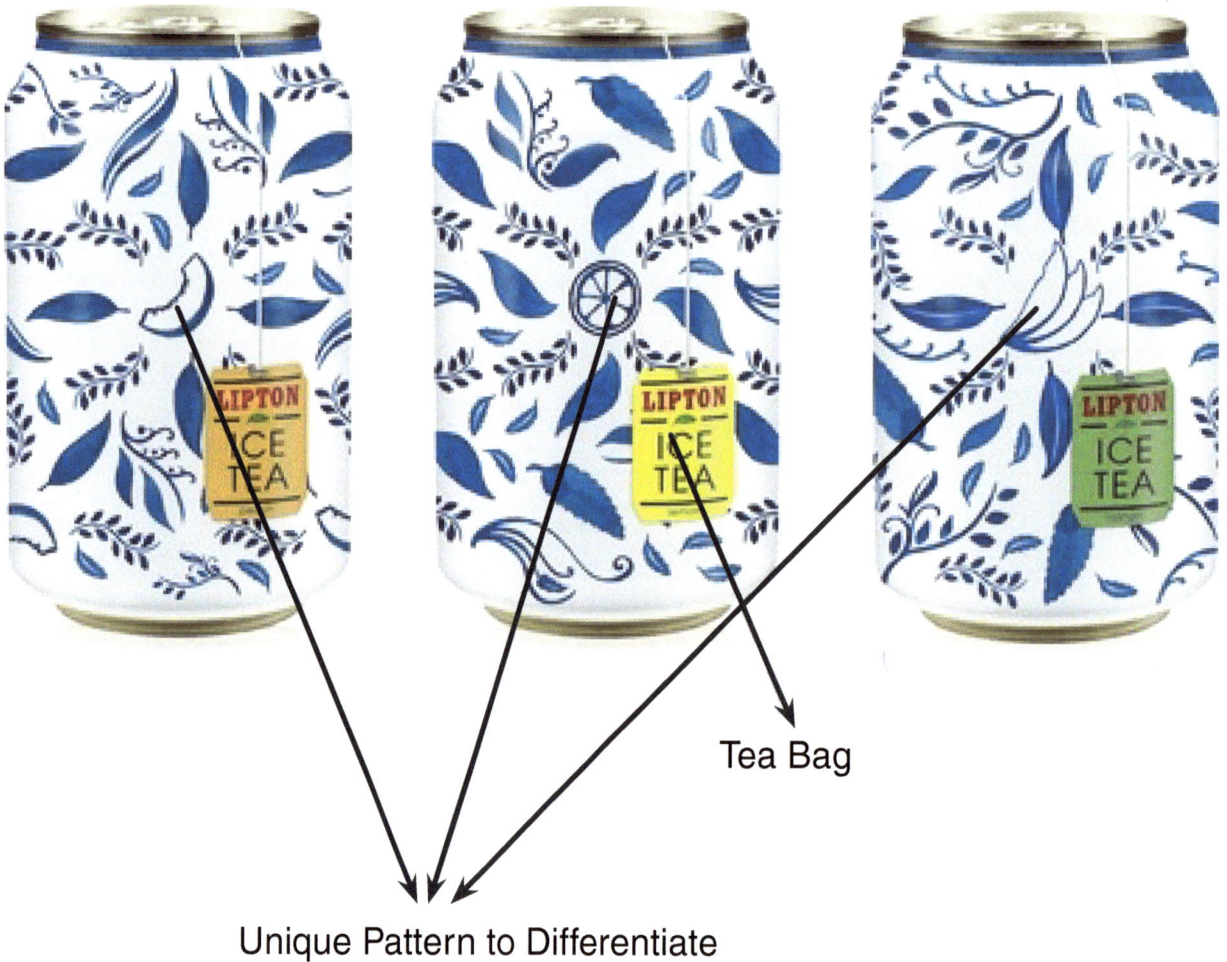

Tea Bag

Unique Pattern to Differentiate

Manufacturer/Designer

Jessica Duncan

- The brand identity is on the teabag tag that hangs over the can.

- Tea bag not only reflect the brand identity but also reflecting the flavours of tea packed because the color of illustrations is same on all the cans.

- Each of the flavoured drinks has same color but a unique pattern to differentiate from Lemon, Peach and Mango.

TEA PACKAGING INNOVATIONS

TENCHA

Green Tea
KAKTUSFEIGE
MALVEN

White Tea
GRANATAPFEL

Rooibos
MANDARINE
JASMIN

TEA PACKAGING INNOVATIONS

TENCHA

Green Tea
KAKTUSFEIGE
MALVEN

TENCHA

White Tea
GRANATAPFEL

TENCHA

Rooibos
MANDARINE
JASMIN

Manufacturer/Designer

ARD Design

- The ice tea in various flavours, which are highlighted by a packaging illustration visualizing the respective fruit and matching coloring.

- A new premium selection of ice teas with an Asian appearance market.

- The design also includes the naming as well as brand and packaging design.

TEA PACKAGING INNOVATIONS

Typography featuring fruit illustrations

Manufacturer/Designer

Wren & Rowe

- Cold tea featured with bright summery colours.

- And oversized typography featuring fruit illustrations.

- Ice tea is packed in tetra pack.

- The impact in-store is immediate, and the range certainly achieves the objective of putting other brands in the shadow.

- The brief is to outshine the brand, and to create a "crowd-pleasing" fun approach to the range.

TEA PACKAGING INNOVATIONS

logo design and its color
Reflecting the product details
like red color for red ice tea

Manufacturer/Designer

Anna Caselli, Italy

- Unique shape of bottle attract maximum attention of consumers.

- Graphic design and color combination reflects the flavours of iced tea.

- The logo design and its color itself reflecting the product packed inside it.

- This iced tea line respects and reinforces the unique values of Twinings, but this time aimed for a younger wealthy audience.

- Target consumers are those who loves taking care of himself, look good and play sports and people who recognize the quality and like to stand out from the crowd.

TEA PACKAGING INNOVATIONS

Manufacturer/Designer

Tea over ice

• This is an innovative concept to prepare fresh chilling ice tea from tea bags with a new container.

• An infuser is designed to prepare cold tea which have unique shape and look like sprouted leaf of tea.

• Infuser and container both are packed in a carton on which infographics are drawn to tell all about the use and procedure.

• Design of infuser and container give authentic and flash chilled moments even before drinking tea.

• We have to just add boiling water over infuser in special container which is placed over iced container and leave it for 3-5 minutes and prepare fresh iced tea.

Manufacturer/Designer

Identity Works

- The concept is based on an idea about the message in a bottle with user-generated messages on the surface of the bottle.

- A peelable label printed with brand name on front and a hidden massage at back.

- This hidden message can be decoded with smart phone for the purpose of counterfeit.

- Background color of label reflects the flavours of tea.

- Use of glass bottle make it more attractive.

- Message in a bottle that idea is probably as old as the bottle itself but this is the new concept of hidden message as well as counterfeit.

TEA PACKAGING INNOVATIONS

Manufacturer/Designer

Maya Kimmel Graphic design & illustration

• Relaxing, healthy (rich in antioxidants), holistic, and serene are the words that best describe the increasingly popular tea beverages.

• This packaging is reflecting this in the design as well as in the materials; recycled and recyclable glass bottles and cardboard.

• So, this packaging is harmless to the environment that's why the target market is health conscious consumers.

TEA PACKAGING INNOVATIONS

Manufacturer/Designer

Harney & Sons

- This packaging has very traditional feeling, enhanced by the Colour choices and style of packaging.

- Color choices not only give traditional but also reflect the flavours of tea

- The cold tea in bottles feel like an odd product for such a traditional tea brand.

- The overall feeling could be a little stuffy and stayed, appealing to a long term customer or those looking for something traditional.

TEA PACKAGING INNOVATIONS

Manufacturer/Designer

Quai Sud

- The glass jar signifies and expensive product.
- The labels applied is simple, chic and understated.
- Different type of cap adding value to glass jar.
- Overall a strong design and very in keeping with the personality of the

TEA PACKAGING INNOVATIONS

DISCLAIMER

TEA PACKAGING INNOVATIONS

Dear Readers!
Thank you for your interest.

We are capable of providing customized reports and surveys.
For further details contact:

amita.venkatesh@packagingconnections.com

CONTACT US

CORPORATE OFFICE

Sanex Packaging Connections Pvt. Ltd.
(ISO 9001:2008 certified company)
117, Suncity Business Tower
Golf course road, Sector--54
Gurgaon-122002
India
Tel: +91 124 4965770
Fax: +91 124 4143951

Email: info@packagingconnections.com
Website:www.PackagingConnections.com